TONY MANGINO

Exploring Objects In Outer Space

*The Complete Guide to Facts and Sizes Of The Most
Mind Blowing Objects In The Universe*

First edition

This book was professionally typeset on Reedsy.
Find out more at reedsy.com

The Earth is the cradle of humanity, but mankind cannot stay in the cradle forever.
— Konstantin Tsiolkovsky

Contents

1

Introduction

Space the final frontier. That is the line from the Star Trek series. Science fiction always aims towards outer space because space is filled with objects of wonder. Man has been amazed with the stars in the sky. ancient Egypt thought the stars were the gods they were worshiping. They even built the pyramids along the constellation of Orain. Later came Galileo, an astronomer that was born in Italy. He discovered that the Earth revolved around the sun which was a big leap in astronomy, a greater understanding of outer space came from that. In 1969 man landed on the moon which made space exploration and a living on another world possible. Humans feel there is something bigger than them in the universe, the ancient Egyptians were right about that. In this book you are going to take a journey to explore the objects of the universe, from the planets of our solar system to far off galaxies. Outer space is vast and full of mysteries that this book will discover. So let's get on our spaceship there no time to waste!

Chapter 1: The Sun - Our Local Star

Our sun is a star, one of the billions of stars in the universe. The sun played a key part of life on planet earth, without the sun the earth will have no light and heat. There would be no rain and no photosynthesis for plants to grow, that means animals don't have any food to eat or water to drink because the water would be frozen due to no heat. With no sun there will be no life intelligence or not. The sun also keeps all the planets in the solar system without the gravity of the sun. All the planets would go off in all directions into deep space. The sun is made up of two gaseous hydrogen and helium however the sun has six layers.

1. radiative zone: The fifth layer of the sun This is where energy is processing into thermal radiation
2. convective zone: The fourth layer of the sun temperatures here are cooler than the rest of the sun however gasses are heated in this zone and shoot out throws outsize zones
3. Photosphere: The third and the lowest visible layer of the sun. This is were the size of the sun is measured

4. Chromosphere: This is the second layer of the sun it's a layer of gas surrounding the photosphere
5. Corona: This is the surface of the sun solar flares and sunspots happen here
6. Core: This is the center of the sun. Temperatures here are 28 million degrees Fahrenheit. In the core nuclear fusion happens.

The size of the sun is so massive that it is about 99.8% of all the mass in the solar system. With that much power solar flares can take place, solar flares are clouds of atoms and electrons that erupt due to sunspots. Sunspots are dark areas of the sun where energy could blow up. The particles from solar flares can affect communication with satellites and can even knock out power grinds on the surface of the earth.

Chapter 2: Planets of the Solar System

- Mercury: The closest planet to the sun it's a little smaller than earth with no atmosphere. The surface of Mercury is hot at 800 degrees Fahrenheit no water on this planet and filled with craters.

- Venus: the second planet from the sun this is the hottest planet in the solar system due to its thick atmosphere that traps heat. Venus is a similar size to earth however Venus surface is filled with volcanoes and temperatures on Venus are hot enough to melt lead.

- Earth: our home planet and the third planet from the sun. Earth has everything life needs: water, oxygen, a protective atmosphere. A right balance of temperatures humans are currently searching for more planets like Earth.

- Mars: The fourth planet from the sun Mars is a dry desert planet with sandstorms. This planet does have ice caps like earth, Mars even used to have water just like Earth. NASA is doing space missions for Mars to learn more about the planet to see if it has life.

- Jupiter: The fifth planet from the sun. Jupiter is the biggest planet in the solar system, so big in fact that hundreds of Earths can fit inside it. Jupiter has no solid surface. It's a gas planet filled with storms. The biggest storm on the planet is called the great red spot. It's been there for centuries. What people may not know about Jupiter is that the planet does have rings but the rings are very thin. Jupiter also has nighty-five moons.

- Saturn: The sixth planet from the sun, a gas giant and known for its rings. Saturn has massive rings, Saturn rings are made up of rock and ice. The rings are created because Saturn's powerful gravity broke up comets and meteoroids.

- Uranus: The seventh planet from the sun. The planet is one of the gas ice giants. The planet is known for its rotated spin at an almost 90 degree angle making it seem like the planet is spinning sideways while orbiting the sun. Uranus does have rings, Uranus has methane gas which gives the planet its blue color. Storms on this planet can reach 560 miles per hour.

- Neptune: Neptune is the eighth and the farest planet in our solar system. Neptune is an ice gas giant planet Neptune is more dense than Uranus, Neptune is made up of the same gasses as Uranus except for an unknown element that makes Neptune a brighter blue color. Neptune winds are three times stronger than Juptier at a scary 1,200 miles per hour. Neptune used to have a storm called the great dark spot; it is now gone and replaced by other storms.

- Pluto and dwarf planets: Pluto was once the ninth planet of the solar system but in 2006 it was reclassified as a dwarf planet. The difference between a planet and a dwarf planet is. A planet is shaped like a circle orbiting around the sun and is clear of objects in its orbit path. A dwarf planet has all those points except there are objects in its orbit path. The solar system has five dwarf planets, Ceres which is in between Mars and Juptier. Pluto which NASA took new pictures of, Haumea which is a fast rotating dwarf planet. Makemake which is the second brightest object in the Kuiper Belt and Eris which is the biggest dwarf planet in the solar system.

Chapter 3: Moons, Rings, and Satellites

S atellites are man made machines that are launched into space using rockets. Depending on how they are created they can have a wide range of uses. There are satellites that study the earth's weather, take pictures of the universe, take pictures of the earth, and provide signals for TV. As you may have seen there are planets that have rings. Rings are parts of ice and rock broken up by a planet's gravity. However these planets have to have powerful gravity in order for rings to form which is why planets like Saturn have them. Moons are bodies in space that orbit planets. For example earth's moon you can see it every night in the sky. The earth moon causes tides on earth so it plays a part in the weather. There are more moons out there that are fascinating to scientists. Let's go to Jupiter where Europa is. Europa has an icy surface and scientists think there is water under the ice and where water is there could be life. NASA is sending a rocket to Europa in 2024 to unlock the key of life it could have under the ice. Next stop is Saturn where Titan is located. Titan is unlike any moon; it has a thick atmosphere with bodies of liquid on the surface. Titan's atmosphere is made up of nitrogen and like Europa scientists think there is water under Titan's icy surface. However get a cold spacesuit because Titan weather can fall below -290 degrees Fahrenheit.

Chapter 4: Asteroids and Meteoroids

An asteroid is a rocky object that orbits the sun; they are also smaller than planets. Most asteroids are in the asteroid belt which is in between Mars and Jupiter. NASA has taken pictures of asteroids for example NASA took pictures of asteroids Mathilde, Gaspra and Ida. Meteorites are parts of asteroids that break off and burn up in the earth's atmosphere. Some meteorites hit the earth surface causing loss of life or just a hole in the ground if lucky. The good news is NASA does have a planetary defense system if an asteroid crosses our path. NASA was successful with its mission in 2022 to move an asteroid out of its orbit by inches, which would be enough to save planet earth from an asteroid.

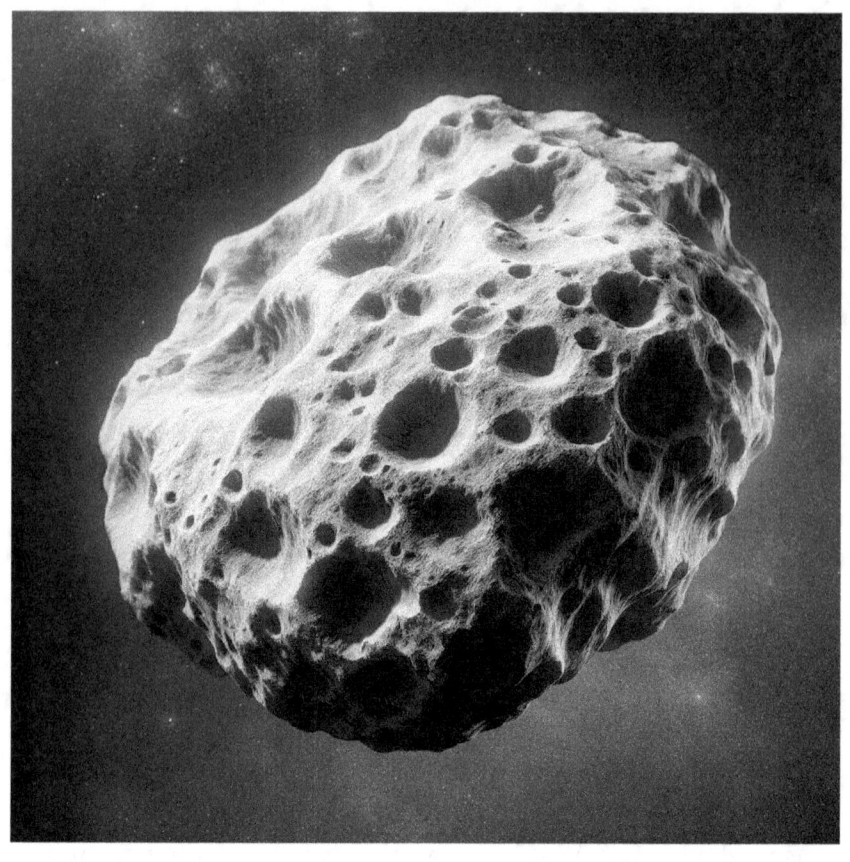

Chapter 5: Comets: The Icy Wanderers

C omets are made of objects that are dust and ice. Unlike the asteroid belt, comets come from the Kuiper belt which is outside Pluto. Comets have tails of light when moving through space. The tail is dust and gas trailing off, the best known comet is Halley's Comet. It's famous because it's been seen by humans for thousands of years. Halley has a 76 year orbit around the sun. Halley was discovered by Edmond Halley where it got its name. A fun fact is Halley produces meteor showers every time it passes earth because of the ice and rock it sprays into space.

Chapter 6: Stars and Constellations

We know what stars are, big balls of heat and gas but the life cycle of a star depends on its mass. All stars start their lives in a nebula. A nebula is a cloud of gas and dust the process begins when a nebula collapses on its gravity. After that a protostar is born. A protostar is the result of a gravity collapse, a protostar can spin at great speeds creating extreme pressure and heat. During this time there is a counterbalance between the star's internal forces and gravity. A protostar can be billions of kilometers in length. The heat and pressure by the star spinning causes the gas cloud to crash in on itself. The process is not over yet the next stage is called T-Tauri. At this stage a star is close to achieve nuclear fusion however temperatures are still low for fusion so in the meantime all the star energy comes from gravitational force. At this stage the star is a size of a small to a medium size star. The stage T-Tauri is very bright; this can last a hundred million years until the star balances out its internal and gravitational forces. That is when nuclear fusion is achieved.

The main sequence is the longest stage of a star's life. A star's life depends on how much mass it has, let's begin with small to medium size stars. They have a lifetime of billions of years. For example our sun is a medium size star. These stars have a red and yellow color. What happens when these stars end the main sequence? Let's find out, once the star has integrated all its hydrogen at its core the nuclear rationdation process stops. Because of this the star collapses in on itself. The star will expand outward from 100 to 1000 its size. On top of that the heat of the star will cool down to 2500 Kelvin. This is called a red giant. The star core will continue to heat to 100 million Kelvin now helium fusion can happen. This will continue until the helium runs out of fuel. The star will explode leaving behind a white dwarf. A small core will remain after the explosion consisting of carbon and oxygen. These white dwarfs are about the size of the Earth and their light will dim until it's no longer a white dwarf but a black dwarf. That is the end of the small to medium sized star life cycle.

For large stars their life cycle is different, during the main sequence they are blue in color and only last a few million years. Large stars turn into red supergiants. The cores of these giants are so high that nuclear fusion can occur after the helium is depleted. Just one red supergiant can cover half a solar system. During nuclear fusion iron is formed, the iron will consume energy in the star until. The moment the star turns to iron. The star will collapse on itself with temperatures up to 100 billion Kelvin. This will trigger the most powerful explosion in the universe called a supernova. Supernovae can be so bright that they can outshine galaxies. During a supernova a fusion of iron occurs and heavy elements are created in the blast. After the blast the outcome of a large star life depends on its mass. If a core of a star has a solar mass of 1.4 to 3, a Neutron star will be formed. Neutron stars are dense and heavy objects incorporated of neutral charged neutrons. For stars with more than 3 solar masses a black hole will be created. Black holes will be covered in a later chapter.

Constellations are groups of stars discovered by humans a long time ago. Take a moment that you are looking at the night sky thousands of years ago. You become perplexed by the stars so much you start drawing and connecting the dots with them in your mind. Constellations are named after objects, animals and people. Back in ancient times constellations were used for Astrology which is a belief that location of planets and stars can determine a person's future. Today constellations are used to locate stars in the universe. Also celestial navigation is using constellations to measure where you are at sea. Despite constellations being a product of the past they are still used for science today.

Chapter 7: Galaxies

G alaxies are a collection of gas, dust, stars, and dark matter. The Milky Way is a galaxy that Earth lives in. Galaxies can have as many as a hundred trillion stars inside them. Galaxies are the best homes for stars, the Milky Way is a spiral galaxy. A spiral galaxy is a rotating object with arms of stars spinning around. Elliptical galaxies contain a lot of old stars and are egg-shaped. Dwarf galaxies are galaxies that have only a few billion stars. Dwarf galaxies orbit bigger galaxies like the Milky Way. The Milky Way has fourteen dwarf galaxies orbiting it, if not more. Irregular galaxies do not fall into the spiral or elliptical categories. There are some theories that irregular galaxies are formed when two galaxies crash into each other. Irregular most likely do not have life in them because newborn stars outnumber the old ones.

Chapter 8: Exotic Objects and Phenomena

B lack holes are formed when a star with more than three solar masses runs off of fuel. Black hole is an area of such intense gravity that nothing can escape from it. The event horizon is the point of no return for anything that gets too close to the black hole death grip. This is where spaghettification happens. Spaghettification is when an object that is falling into a black hole is stretched out and compressed until it falls into the singularity at the center of a black hole. The singularity is matter that has collapsed into an infinite dense small circle. Although there are some theories that explain that on the other side of black holes there are parallel universes, the singularity is the most agreed upon. Supermassive black holes are black holes millions of times or more bigger than the mass of the sun. No one knows how they were created but there is a theory that many smaller black holes meet with each other to create a supermassive black hole. Supermassive black holes can be found in the center of galaxies which many scientists think are why stars spin around it. This may sound ridiculous but a supermassive black hole may be the reason why there is a Milky Way.

A pulsar is when a large star reaches the end of its life and creates a neutron star. A pulsar is just another term for this, a quasar is a type of galaxy. Quasars are young galaxies still forming even though both of the names sound like types of stars that are not the case, that is why they are confused with each other.

Dark energy is the name given to the phenomena of the increased speed of expansion of the universe. Dark energy is a big mystery in the universe. Scientists know it's there but can't detect it. However dark energy does make up 68% of the universe. 27% of the universe is made up of dark matter. Dark matter is matter we can not see with our eyes. Scientists think one of the ingredients of dark matter is the particle baryon. Scientists know about the baryon particle because they are able to detect baryon clouds through radiation. There is another theory that dark matter is made up of WIMPS or weakly interacting massive particles. Whatever dark matter is, it would be the one of the most important discoveries of all time.

Chapter 9: The Search for Extraterrestrial Life

Mankind has asked the age old question for a long time, are we alone in the universe? With billions of galaxies and even more stars the possibility of us being alone is not zero. The search for life begins in our own backyard. We know that one of the keys for life is water. Mars is a planet with a similar size compared to Earth. It's a rocky planet with ice caps, where there is ice there is water. NASA has sent rovers Curiosity and Perseverance to study Mars. Scientists know there is water under the ice on Mars.The question is how do you get under the ice, future missions to the red planets may solve that question. A rover can drill under the surface on the ice to find water. Water may flow on the top of the ice when the planet is closest to the sun. Exoplanets are planets that are outside our solar system. The first exoplanet discovered was in fact two planets. Poltergeist and Phobetor are planets found orbiting the pulsar PSR B1257+12 b. The two planets have no water on them because of the pulsar's intense radiation. Despite that, the discovery confirmed that there are more planets out there. 5,572 exoplanets have been confirmed by NASA as of the time of writing this book. NASA has found planets in the habitable zone. The habitable zone is the distance a planet has to be from a star in order for life to form. The planet Earth is in the habitable zone, if Earth was too close to the sun the Earth would be covered with lava. If the Earth was too far away the Earth would be too cold. The Trappist-1 solar system has seven rocky planets. With two planets in the habitable zone, it is a great discovery that this will be the solar system where we are most likely to find life. With how big the universe is, the number of exoplanets are endless.

Chapter 10: The Age and the Size of The Universe

The big bang theory is the leading creation story of the universe. There is some confusion with the big bang, it's not an explosion that came out of nothing. Rather at the beginning of the universe there was a dense singularity of space. One day, if you want to call it that, an explosive expansion began. There is another theory on why the explosion happened. Two parallel universes come in contact with each other creating a big bang. There is also another theory that explains that when a universe comes back down to a big crunch when that universe dies on the other side will be a big bang. There is strong evidence to support the big bang theory because of mathematical equations and cosmic radiation. NASA has a map of this cosmic radiation, the radiation was discovered by using a radio receiver. The radiation can help scientists study the age of the universe. The Planck satellite studied the background radiation and the data that came back was that the universe was 13.82 billion years old. A little more than one billion years earlier than previously thought. The current size of the universe is 94 billion light years. Take a moment on how big that is, the speed of light is 186,282 miles per second. However add those miles up to a year, the universe is so big that we may have to invent new numbers to fathom it. The farthest object that the Hubble telescope can see is a galaxy 13.2 billion light years away. Based on the measurements that light would take 13.2 billion years to get to Earth. So we are seeing 13.2 billion years in the past.

12

Conclusion

We started our journey with the sun, the life force that gives us light. Then we traveled to the solar system to visit the planets and moons. Some may have life waiting to be found. However asteroids ram though the solar system not caring who they hit. The good news is we have plans if one gets too close. We learned that the stars give us life and there are new ones being discovered. Galaxies are home to solar systems. Black holes are a dark force. We may not be alone in the Milky Way and the beginning started with a big bang. Despite everything we know there are still more mysteries that have to be solved. What is dark matter? What is dark energy? Are there more particles that build the universe? Do aliens exist out in outer space and if so are they wondering the same thing? Those questions could be answered if we support space exploration. With the moon landing the world came together, this was during the time when the cold war was raging between the United States and Soviet Union. The world was at peace during the Apollo moon landing, space really bought mankind together for infinity and beyond. However we still need to keep that going. More money should be spent on space exploration because someday the Earth will no longer be here. If mankind wants to continue finding another planet to live on is a goal and one that the Apollo astronauts would want for us.

13

Resources

Imagine the universe! (n.d.). https://imagine.gsfc.nasa.gov/feature s/cosmic/farthest_info.html

Van Helden, A. (2024, January 23). *Galileo | Biography, Discoveries, Inventions, & Facts*. Encyclopedia Britannica. httpsSun. (n.d.).

Sun. (n.d.). https://education.nationalgeographic.org/resource/sun /

Venus: Facts - NASA Science. (n.d.). https://science.nasa.gov/venus/facts/

Mercury: Facts - NASA Science. (n.d.). https://science.nasa.gov/merc ury/facts/

Jupiter: Facts - NASA Science. (n.d.). https://science.nasa.gov/jupiter/

Saturn: Facts - NASA Science. (n.d.). https://science.nasa.gov/saturn/f acts/

Uranus: Facts - NASA Science. (n.d.). https://science.nasa.gov/uranus/ facts/

Pluto & Dwarf Planets - NASA Science. (n.d.). https://science.nasa.gov /dwarf-planets/

Neptune: Facts - NASA Science. (n.d.). https://science.nasa.gov/neptu ne/facts/

Moons - NASA Science. (n.d.). https://science.nasa.gov/solar-system/

moons/

What is an asteroid? | NASA Space Place – NASA Science for Kids. (n.d.). https://spaceplace.nasa.gov/asteroid/en/

What is a comet? | NASA Space Place – NASA Science for Kids. (n.d.). https://spaceplace.nasa.gov/comets/en/

1P/Halley - NASA Science. (n.d.). https://science.nasa.gov/solar-syste m/comets/1p-halley/

Staff. (2023, February 2). *Life Cycle of a Star: stages, facts, and diagrams.* Science Facts. https://www.sciencefacts.net/life-cycle-of-a-star.html

What are constellations? | NASA Space Place – NASA Science for Kids. (n.d.). https://spaceplace.nasa.gov/constellations/en/

information@eso.org. (n.d.). *Galaxy.* ESA/Hubble | ESA/Hubble. https://esahubble.org/wordbank/galaxy/

Ktmoelle. (n.d.). *Irregular galaxies.* Ask an Earth and Space Scientist. https://askanearthspacescientist.asu.edu/irregular-galaxies

Tillman, N. T., & Dobrijevic, D. (2023, May 19). *Black holes: Everything you need to know.* Space.com. https://www.space.com/15421-black-ho les-facts-formation-discovery-sdcmp.html#section-types-of-black-hol es

Universe Guide. (2024, January 29). *What is the difference between a Pulsar and a Quasar?* https://www.universeguide.com/blogarticle/wha t-is-the-difference-between-a-pulsar-and-a-quasar

Exoplanet Exploration: Planets Beyond our Solar System. (n.d.). Exo-planet Exploration: Planets Beyond Our Solar System. https://exoplan ets.nasa.gov/

Howell, E., & May, A. (2023, July 26). *What is the Big Bang Theory?* Space.com. https://www.space.com/25126-big-bang-theory.html

Cosmic Times. (n.d.). https://imagine.gsfc.nasa.gov/educators/progra ms/cosmictimes/educators/guide/age_size.html

www.ingramcontent.com/pod-product-compliance
Lightning Source LLC
Chambersburg PA
CBHW071018290526
45795CB00005B/1857